REPTILE GROUPS

BY SUE BRADFORD EDWARDS

Core Library

An Imprint of Abdo Publishing
abdobooks.com

Cover image: Some types of turtles, such as the red-eared slider, bask in the sun in groups. Sometimes they sit on top of each other while basking.

abdobooks.com

Published by Abdo Publishing, a division of ABDO, PO Box 398166, Minneapolis, Minnesota 55439.
Copyright © 2026 by Abdo Consulting Group, Inc. International copyrights reserved in all countries.
No part of this book may be reproduced in any form without written permission from the publisher.
Core Library™ is a trademark and logo of Abdo Publishing.

Printed in the United States of America, North Mankato, Minnesota.
052025
092025

Cover Photo: Anna S. Mariola/Shutterstock Images
Interior Photos: Ken Griffiths/Shutterstock Images, 4–5; Shutterstock Images, 7, 9, 12, 19, 20, 31, 32, 40, 43; Kryssia Campos/Moment/Getty Images, 11; Jim Schwabel/Shutterstock Images, 14–15; Marc Pletcher/Shutterstock Images, 17; Wildlife GmbH/Alamy, 22; Natalia Kuzmina/Shutterstock Images, 24–25; Ken Gillespie Photography/Alamy, 27; Daniel Baleckaitis/Alamy, 29, 45; J. Gerard Sidaner/Science Source, 34–35; Tarcisio Schnaider/Shutterstock Images, 36; Alex Mustard/Nature Picture Library/Alamy, 38

Editor: Laura Stickney
Series Designer: Ryan Gale

Library of Congress Control Number: 2024948984

Publisher's Cataloging-in-Publication Data

Names: Edwards, Sue Bradford, author.
Title: Reptile groups / by Sue Bradford Edwards
Description: Minneapolis, Minnesota: Abdo Publishing, 2026 | Series: Strength in numbers: animal groups | Includes online resources and index.
Identifiers: ISBN 9781098297299 (lib. bdg.) | ISBN 9798384919810 (ebook)
Subjects: LCSH: Reptiles--Juvenile literature. | Reptile populations--Juvenile literature. | Animal colonies--Juvenile literature. | Reptiles--Behavior--Juvenile literature.
Classification: DDC 597.9--dc23

CONTENTS

THE GREAT DESERT SKINK

As the sun sets over the Australian desert, six lizards wait near the entrance to an underground termite colony. When a termite emerges, one of the lizards eats it. Termites are social insects that eat plant-based materials such as leaf litter and wood. They venture out of their colonies to harvest grass. The lizards know where to find these termites, and they continue to feed on them until the sun sets. Then the lizards head back to their burrow.

The great desert skink is also known as the Tjakura. Groups of these lizards can be found in Australia's Central Desert region, including in Uluṟu-Kata Tjuṯa National Park.

BLACK ROCK SKINKS

The great desert skink isn't the only lizard that lives in family groups. In 2003, the journal *Molecular Ecology* published a report written by scientists from the University of Sydney in Australia. After studying 115 Australian black rock skinks, the scientists made some new discoveries. Earlier scientists had noted that these lizards live in groups. But when the university scientists sampled the DNA of each lizard in a group, they discovered the groups consisted of parents and their young. Young skinks stayed with their parents for a little more than a year. During this time, scientists think the adult skinks protect the young.

The lizards are great desert skinks. These reptiles can grow up to about 16 inches (41 cm) long. They are covered with smooth scales. Their backs, heads, and legs are brownish orange, while their undersides are creamy white or light grey. The skinks' tails are longer than their bodies, and they have rounded snouts.

As night falls, the skinks enter their burrow and follow its underground tunnels. They live in a family

GREAT DESERT SKINK BURROWS

This graphic shows a great desert skink burrow, which is made of many interconnected tunnels. What did you think the skinks' burrow looked like before seeing this diagram?

group that consists of a male, a female, and their young. The Australian desert where the skinks live is a land of extreme temperatures. During the day, it can reach 120 degrees Fahrenheit (49°C). At night, the temperature drops significantly. Like other reptiles, skinks cannot warm or cool their bodies internally, which means that they must use their environment to do so. Together, the skinks dug a burrow in the ground that reaches a depth where the temperature is stable. They maintain their burrow in order to survive underground.

Inside the burrow, the skinks head in different directions. The youngest skinks have opened a new entrance that only they can use. Their older siblings and parents are too large to fit through it. Before the family settles down to sleep, the adult male skink heads down a tunnel and goes back outside. He poops in the outdoor latrine area, which all the skinks in the burrow share. Then he hurries back underground to the more comfortable temperatures inside the burrow. The family's system of tunnels stretches across an area that is 40 feet (12 m) wide with 20 different entrances.

REPTILES LIVING TOGETHER

Great desert skinks are one of many reptiles that live in groups. Being part of a group helps animals survive in many ways. Sometimes it provides safety

Marine iguanas form groups to stay warm when the weather is cold. They huddle together and sometimes sleep piled on top of one another.

from predators. Predators will often go after animals that are slow, vulnerable, or alone. Because there are so many animals in a group, some group members will survive. Other times, being in a group helps parents protect their eggs or raise their young. Groups also allow animals to share resources that they need to survive. By working together, animals can maintain a home more easily too.

Reptiles are a group of animals that have scales or bony plates covering their skin. They breathe air. They are ectothermic, which means they cannot keep themselves warm internally. Instead, they get warmth from the sun. Reptiles include snakes, lizards, turtles, and crocodilians.

Snakes are legless animals. They have no movable eyelids, so they cannot blink. Snakes have no ear openings either. They use their forked tongues to pull air particles into their mouths, which helps them smell. Different types of snakes have different group names.

DESERT NIGHT LIZARDS

In the 1960s, scientists found groups of desert night lizards under Joshua tree logs in the Mojave Desert in California. Biologist Alison Davis, a researcher at the University of California, Berkeley, wanted to see these lizards for herself. She noticed something about the lizard groups she found. They often contained an adult male, an adult female, and young lizards of various ages. These were usually a mother and father lizard and their offspring.

A group of cobras, for example, is called a quiver. A group of rattlesnakes is known as a rhumba.

Some lizards, such as skinks, have smooth scales. Others, such as iguanas, have pointed scales along their spines. Most lizards have movable eyelids that can blink. There are 4,000 lizard species around the world. Any group of lizards is called a lounge, but a group of iguanas can also be called a mess.

Turtles have shells that protect their backs and stomachs. These shells are made of bones covered in large scales called scutes. A group of turtles is known as a bale.

Crocodiles, alligators, and gharials are all crocodilians. Crocodilians have long, narrow bodies. Their heavy scales form protective armor. They also have strong tails, which they use to swim. A group of crocodiles is called a bask.

Groups of reptiles live together for specific reasons. Parent–offspring groups allow crocodilians and lizards to raise and protect their young. Reptiles may also form groups when looking for shelter from the cold. Scientists have observed social behavior in some snakes and turtles but are still learning what this behavior means. No matter the size, groups help many reptiles survive and thrive in the wild.

FURTHER EVIDENCE

Chapter One of this book discusses the great desert skink. What was one of the main points of this chapter? What evidence is included to support this point? Read the article at the website below. Does the information on the website support the main point of the chapter? Does it present new evidence?

GREAT DESERT SKINK

abdocorelibrary.com/reptile-groups

CROCODILIANS

American alligators are crocodilians that live in swamps, rivers, and ponds. Their range extends from North Carolina to Texas and Oklahoma. They live only in the United States, farther north than any other crocodilian. These reptiles can grow to be 12 feet (3.7 m) long and weigh up to 1,000 pounds (450 kg). Their dark coloring helps them blend in with the water, while their long tails and webbed feet propel them forward as they swim.

American alligators have long, rounded snouts with nostrils that face upward. They can have up to 80 teeth in their mouth at a time.

15

After a male and female alligator mate, the female returns to her territory. She makes a nest there. She builds this nest out of whatever plant matter is in the area. It can take up to an hour for the female alligator to lay an average of 39 eggs. When she is done, she covers the eggs with plant matter. As this plant matter rots, heat is produced. This warms the eggs and helps them mature.

Heat also determines the sex of the babies. A temperature of 88 degrees Fahrenheit (31°C) or below produces only female babies, while a temperature of 91 degrees Fahrenheit (33°C) or higher produces only males. Temperatures in between 88 and 91 degrees Fahrenheit produce a mix of males and females.

The female alligator doesn't sit on her eggs like a bird would. If she did, she would crush the eggs. Instead, she stays within about ten feet (3 m) of the nest to protect the eggs from predators. These include raccoons, black bears, opossums, river otters, and crows.

American alligator babies sometimes ride on their mother's back while she swims through the water. This helps the mother protect the babies from predators.

It takes about 65 days for the baby alligators to mature inside the eggs. When they are ready to hatch, the babies make high-pitched noises. The mother removes the nesting material to uncover the hatching eggs. The young alligators are about six to eight inches (15–20 cm) long when they hatch.

Newly hatched American alligators live in small groups called pods. Up to 80 percent of the hatchlings fall prey to birds, raccoons, bobcats, snakes, bass, and other alligators. This rate would be even higher if the mother alligator did not defend her young, driving off potential predators.

NILE CROCODILES

Alligators aren't the only crocodilians that form parent–offspring groups. An adult Nile crocodile can grow up to 16.4 feet (5 m) long and weigh as much as 1,300 pounds (590 kg). Nile crocodiles are larger than alligators and have narrower, pointier snouts. They live in a variety of aquatic habitats throughout eastern Africa, including freshwater lakes, rivers, and swamps. They can

GHARIALS

Gharials are crocodilians that live in India. Many observers have seen only female gharials defending young. An organization called the Gharial Conservation Alliance fitted gharials with radio transmitters. This made it possible to locate the animals in the wild and track their movements. In this way, researchers were able to observe individual gharials. Members of the organization discovered that some large males also defend young gharials from predators.

also be found where rivers meet the ocean and in mangrove swamps. Mangroves are tropical trees that grow in water.

After a male and female Nile crocodile mate, the female selects a location that is close to the water's edge. Then she digs a hole that is 16 to 20 inches (41–51 cm) deep. In this nest, she lays 25 to 90 eggs. Then she covers the eggs with sand. Like female American alligators, female Nile crocodiles chase predators away from their nests. Although few animals are large enough to threaten an adult Nile crocodile,

some eat crocodile eggs. These include baboons,
mongooses, and a lizard called the Nile monitor. It is
up to the female crocodile to keep her eggs safe from
these predators.

After 90 days, the eggs are ready to hatch. Like
alligator eggs, the sex of developing crocodiles is
determined by the temperature of the nest. If the
temperature is below 87 degrees Fahrenheit (31°C),

the eggs will develop into females. Males will develop if the temperature is between 87 and 93 degrees Fahrenheit (31 and 34°C). When the eggs have matured, the babies chirp to get their mother's attention. Then the mother uncovers the nest. If the nest is too far from the water, the female crocodile picks up the babies in her mouth and carries them to the water.

The amount of time that the mother spends with her hatchlings depends on how much food is available in the area. If food is plentiful, the young crocodiles will be able to find their own food.

Nile crocodiles have the strongest bite of any animal on earth. But they are able to gently carry their babies and hold their eggs in their mouths.

The mother will stay in the area for several months to protect them. But if there isn't much food in the area, the mother will leave. This encourages the young crocodiles to go elsewhere, seeking safety and food. By protecting the nest and the young for as long as possible, the female Nile crocodile forms a parent–offspring group with her young. This helps ensure their survival.

STRAIGHT TO THE
SOURCE

The Croc Docs are a team of researchers from the University of Florida. They study crocodilians. On their website, the team discusses American crocodile nests in Florida:

Although the majority of American crocodiles in Florida Bay deposit their eggs within holes, some individuals build mound nests. In Florida Bay, the majority of nests are hole nests, with a few mound nests located mainly on islands. . . . Raccoons are the dominant natural predator of crocodile nests in Florida and are responsible for a few to many nest failures each year. A female crocodile constructs a maximum of one nest per year and can nest as often as every year or every couple of years. . . . She can build her nest away from others, or a single site can contain several nests [close] to each other. . . . Unlike with sea turtles, hatchling crocodiles cannot dig their way out of the nest. They are dependent on mom to open the nest up.

Source: "Nesting Crocodiles and Hatchlings." *Croc Docs*, n.d., crocdoc.ifas.ufl.edu. Accessed 7 Jan. 2025.

WHAT'S THE BIG IDEA?

Take a close look at this passage. What is the main connection being made between American crocodile nests and hatchlings? How do mother crocodiles protect their nests?

GARTER SNAKES

Garter snakes can be found throughout North America and Central America. Different species have different colorings. Common garter snakes live throughout the United States and in southern Canada. They are dark in color, with many featuring shades of black, brown, or green. They often have three yellow or white stripes. One stripe runs down the snake's spine, and one runs down each of its sides.

Common garter snakes can be found in wet, grassy areas such as gardens and wetlands. They frequently hide or make dens under rocks.

The largest garter snakes can grow up to 34 inches (86 cm) long, although most are shorter.

Garter snakes benefit from living in several kinds of groups. These include large numbers of snakes that can be found together in hibernacula. Hibernacula are communal dens in which many snakes gather to spend the winter. Most types of garter snakes form these dens. Near the town of Narcisse in Manitoba, Canada, visitors come to see hibernacula called the Narcisse Snake Dens.

The Narcisse Snake Dens are beneath a series of limestone outcroppings. Limestone is a soft stone that breaks down easily. Red-sided garter snakes follow the stone's cracks and crevasses down below the frost line, which is the depth below which the ground does not freeze in winter. They spend the winter there. Scientists estimate that a single den can hold as many as 10,000 garter snakes. By clustering together, the snakes retain more body heat than a single snake can. This helps them survive until warm weather arrives.

There are four dens in use at the Narcisse Snake Dens. Visitors can see red-sided garter snakes at the site in the spring and fall.

While these snake dens are called hibernacula, reptiles do not hibernate in the same way some mammals do. During hibernation, an animal's body temperature drops. So does its breathing and heart rate. This lasts throughout the winter regardless of external temperature changes.

But reptiles and amphibians brumate. This also involves slowing their breathing and heart rate. But if temperatures warm up enough, a brumating animal can find a patch of sunlight to bask in, which raises its body temperature. This makes its breathing and heart rate speed up. The animal can come out of brumation at any point in time if temperatures warm up enough. The process helps it save energy and survive through cold weather.

THE MATING BALL

When brumation ends and garter snakes emerge from their dens in the spring, it is almost time to mate. This leads to a different

type of group formation. About 14 days after emerging from the den, a female garter snake signals that she is ready to mate by giving off chemicals. Male snakes pick up this chemical scent and rush to find the female. If more than one male finds her, they try to push each other out of the way. The result is a writhing mass of garter snakes called a mating ball. Only the strongest males succeed in mating with the female.

Young garter snakes grow in the female's lower abdomen. The young are born live two or three months later. Female garter snakes usually give birth to between 10 and 40 babies. Larger female snakes have larger litters. When they are born, baby garter snakes are

ARIZONA BLACK RATTLESNAKES

Like some other snakes, the Arizona black rattlesnake stays near its young. When babies are younger than three days old, they aren't very cautious. The young snakes often move too far away from the den. The mother snake herds them back toward the safety of the den. If a predator approaches the den, the mother rattles her tail. She coils her body and raises her head off the ground. She puffs up and tries to look big. These behaviors are threatening to the predator. The mother also keeps the babies warm. Because adult rattlesnakes are larger than babies, the adults retain heat longer. The mother lets young snakes gather on top of her to share heat.

independent and find their own food.

SOCIAL BEHAVIORS

Morgan Skinner is a behavioral ecologist who studied at Wilfrid Laurier University in Waterloo, Ontario, Canada. Behavioral ecologists study animal behavior. Skinner studied captive garter snakes.

In 2020, Skinner placed 10 snakes in an enclosure that was almost 11 square feet (1 sq m) in size. It had

four small shelters in which snakes could hide. Twice

a day, the snakes were photographed and removed

from the shelter. Then they were returned to the

enclosure. The snakes repeatedly sheltered in groups,

often with the same individuals. Skinner wondered

whether wild snakes would show the same behavior.

Then he learned that the Ontario Ministry

of Transportation had studied more than 3,000

The Ontario Ministry of Transportation studied a species called the Butler's garter snake. It looks similar to the common garter snake.

garter snakes. The snakes had to be moved because of road construction, and the Ministry of Transportation wanted to make sure the reptiles thrived after the move. For 12 years, they had repeatedly captured the snakes to check on them. During each capture, they noted where individual snakes were found.

When Skinner and other researchers looked at the data, they saw that the same snakes were frequently found together. On average, these groups included only three or four snakes. But some included up to 46 snakes. The researchers also realized that an older female snake was at the center of many groups. Younger snakes often followed the older snake to wherever she chose to take shelter. Researchers know that being part of a group can help snakes stay warm. Many think that living in groups may also make it easier for garter snakes to spot predators and find food.

TURTLES

Scientists used to think that turtles were not social reptiles. If turtles gathered on a sunny log above the water, scientists believed it was simply because there was only one good log to gather on. But research suggests something different.

In 2020, a pair of scientists paddled a canoe down a muddy river in Belize. This is a country in Central America. The scientists put a hydrophone in the water. This is a device similar to an underwater microphone.

The Central American river turtle is also known as the hicatee. It is one of the largest freshwater turtle species in Central America.

Like Central American river turtles, giant South American river turtles often swim together. These reptiles also migrate and nest in groups.

They listened for sounds from transmitters attached to the shells of Central American river turtles. By tracking these sounds, the scientists were able to tell that the turtles were moving around in groups.

The scientists wanted to know why the turtles were moving together. To learn more, they studied a stretch of river that had no food, logs, or rocks that might attract turtles to a certain area. Even in these areas, the turtles still moved in groups.

The scientists aren't sure why these river turtles travel in groups. But one possibility may be that groups provide safety from crocodiles. The scientists hope their findings encourage other researchers to look more closely at the behavior of turtles and other reptiles.

HAWKSBILL SEA TURTLES

Some sea turtles, such as hawksbill sea turtles, are also social reptiles. At first, scientists didn't believe that sea turtles were social animals. But then they saw a video of hawksbill sea turtles made by a naturalist off the coast of Hawaii. It can be difficult to observe sea turtles in the wild because they are shy and usually avoid humans. But sea turtles have been a protected species in Hawaii for a long time.

PANCAKE TORTOISES

Pancake tortoises live in the African countries of Kenya and Tanzania. These reptiles have flat, flexible shells that allow them to squeeze into crevices to escape heat and predators. Pancake tortoises often live alone. But when they seek shelter from the heat, a pair of tortoises may shelter together. As many as ten pancake tortoises have been found in a single crevice.

Hawksbill sea turtles usually come together only to mate. But some people have observed the animals touching their heads and beaks together. Researchers think these interactions may help turtles identify one another.

This means people are not allowed to hunt turtles, and many turtles no longer fear humans.

A student from Arizona State University watched the naturalist's videos. In the videos, three adult female hawksbill sea turtles and one young hawksbill sea turtle interacted 149 times. The interactions included the sea turtles rubbing their faces together and swatting at one another. They were seen biting and chasing one another too.

The videos proved that hawksbill sea turtles interact with one another. However, scientists are still unsure why. They don't know whether the turtles travel together all the time or whether they come together only for short periods of time. If sea turtles spent longer periods of time together, it could put them at risk of being caught or being hit by a boat. Scientists continue to study hawksbill sea turtle groups to understand how to keep the species safe.

For reptiles, social groups serve

EUROPEAN POND TURTLES

When scientists study animal groups, they look for social structures. These could be cooperative groups. Or they might involve a hierarchy in which one turtle is ranked higher than the others. Between 2017 and 2018, scientists from the University of Milano-Bicocca in Italy studied 16 European pond turtle hatchlings. They discovered that among the hatchlings, some turtles were dominant and bit the other turtles. When a turtle is dominant, it is more powerful than others. The dominant turtles were not always the biggest ones.

PARTS OF A SEA TURTLE

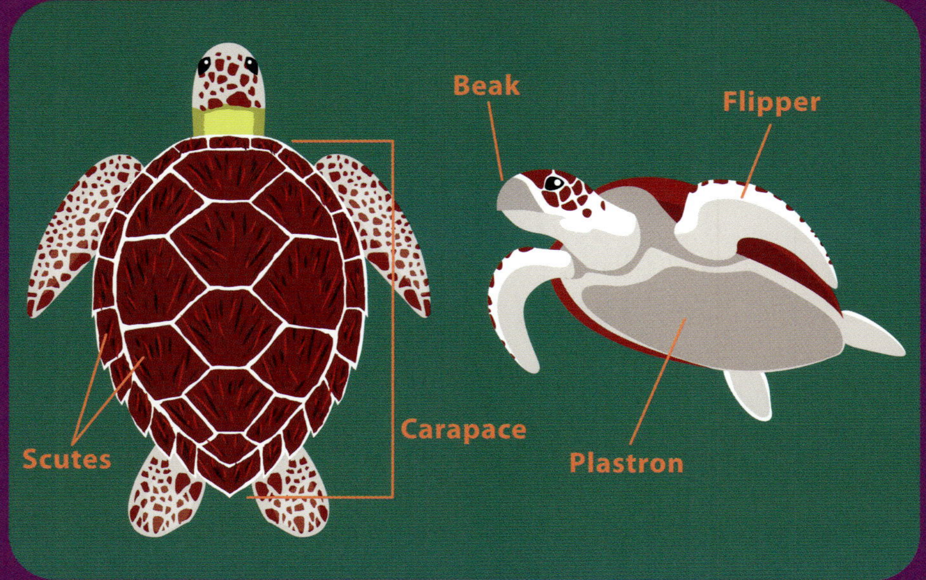

This diagram shows the parts of a hawksbill sea turtle's body, including its shell, which is part of its skeleton. The shell is made up of two main pieces, the carapace and the plastron. What do you notice about the turtle's shell? How do you think the shell protects the turtle?

many purposes. Groups of lizards maintain complicated burrows, while crocodilians protect their nests and hatchlings from predators. Snakes keep their young safe and warm until they learn to care for themselves. Turtles interact with one another. Living in groups helps all kinds of reptiles survive and thrive in many habitats around the world.

STRAIGHT TO THE
SOURCE

Jesse Senko is an assistant professor at the University of Arizona. He spoke about the group behavior of hawksbill sea turtles:

[The head touching] sometimes lasted several minutes and involved turtles rubbing the sides of their faces together, swiping their beaks in a [circular] motion, or pumping water in and out of their mouths and noses. The latter behavior pushes water past their chemosensory organs, which may allow the turtles to smell and thus recognize each other. . . . This study shows that we are just beginning to understand these animals, and that they are likely more complex than we previously realized or assumed. . . . This has important conservation implications for all sea turtles, but especially hawksbills, which are critically endangered and hunted for their shells in several locations worldwide.

Source: Scott Seckel. "Snuggling Sea Turtles." *ASU News*, 26 Oct. 2021, news.asu.edu. Accessed 31 Oct. 2024.

CONSIDER YOUR AUDIENCE

Adapt this passage for a different audience, such as your principal or friends. Write a blog post conveying this same information for the new audience. How does your post differ from the original text and why?

FAST FACTS

- Reptiles are a group of ectothermic animals. They get warmth from their environment. Reptiles include lizards, crocodilians, snakes, and turtles.

- Great desert skinks live in family groups that maintain underground burrows. These burrows protect them from extreme desert temperatures.

- Female American alligators guard their nests to keep their eggs safe from predators. Hatchlings live in pods.

- Female Nile crocodiles guard their nests. They gently crack their eggs open when it is time for the young to hatch.

- Many types of garter snakes shelter together in dens to stay warm during the winter. There may be as many as 10,000 snakes in one den.

- Garter snakes form mating balls. Male snakes pick up the chemical scent of a female and rush to find her. If more than one male finds the female, they form a writhing mass to try to push each other out of the way.

- Central American river turtles have been observed traveling in groups. This may help protect them from predators such as crocodiles.

- Hawksbill sea turtles interact with one another. They rub their faces together, swat at each other, and chase each other. Scientists are still learning why hawksbill turtles behave like this.

Surprise Me

Chapter Two discusses crocodilian behavior. After reading this book, what two or three facts about crocodilians did you find most surprising? Write a few sentences about each fact. Why did you find each fact surprising?

Dig Deeper

After reading this book, what questions do you still have about reptile groups? With an adult's help, find a few reliable sources that can help you answer your questions. Write a paragraph about what you learned.

Another View

This book talks about garter snakes. As you know, every source is different. Ask a librarian or another adult to help you find another source about these snakes. Write a short essay comparing and contrasting the new source's point of view with that of this book's author. What is the point of view of each author? How are they similar and why? How are they different and why?

Say What?

Studying reptile behavior can mean learning a lot of new vocabulary. Find five words in this book you've never heard before. Use a dictionary to find out what they mean. Then write the meanings in your own words and use each word in a new sentence.

GLOSSARY

captive
describing a wild animal that has been caught and kept by people

chemosensory organ
an organ in an animal's body that detects chemicals produced by another animal, often of the same species

communal
shared or used by multiple individuals

conservation
the preservation and protection of a wild animal or natural environment

DNA
deoxyribonucleic acid, a chemical that is the basis of genetics, through which various traits are passed from parents to offspring

mature
to develop or grow

migrate
to move from one place or region to another

particle
a small piece of something

range
the geographic area in which an animal lives

vulnerable
in danger of being harmed

ONLINE RESOURCES

To learn more about reptile groups, visit our free resource websites below.

Visit **abdocorelibrary.com** or scan this QR code for free Common Core resources for teachers and students, including vetted activities, multimedia, and booklinks, for deeper subject comprehension.

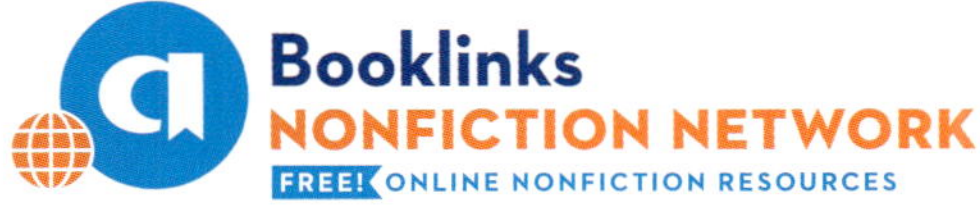

Visit **abdobooklinks.com** or scan this QR code for free additional online weblinks for further learning. These links are routinely monitored and updated to provide the most current information available.

LEARN MORE

Russo, Kristin J. *Reptiles*. Abdo, 2024.

Somaweera, Ruchira. *The Ultimate Book of Reptiles*. National Geographic, 2023.

INDEX

About the Author

Sue Bradford Edwards grew up in Missouri watching every PBS special on animals that she could find. She cared for both Texas horned lizards and a green iguana. She is now a Missouri nonfiction author who writes about science, social sciences, and culture. She has written more than 60 books for young readers.